TASMANIAN MAMMALS

A

FIELD GUIDE

TASMANIAN MAMMALS

A FIELD GUIDE

PHOTOGRAPHS BY

DAVE WATTS

TASMANIAN CONSERVATION TRUST

Photographs by Dave Watts

Text compiled by Margaret Grainger, Emma Gunn, and Dave Watts.

Design by Ian Skinner.
Set in Berkeley book by Type Faces, Hobart

Printed by Focal Printing, Hobart.

Published by the Tasmanian Conservation Trust Inc.
102 Bathurst Street
HOBART TASMANIA 7000.

ISBN 0 9590500 4 3

ACKNOWLEDGEMENTS

Many people have given freely of their time to advise on the text and to help ensure the accuracy of the information.

We are most grateful to David Rounsevell and Greg Hocking of the Department of Lands, Parks and Wildlife, for their continued help with this project.

Dr. Robert Taylor and Nick Savva deserve special thanks for their expert knowledge and help with the difficulty of identifying bats.

Peter Brown, Philip Hoysted, Peter Blackwell and Debbie Quarmby gave much needed support and constructive criticism to the text.

We appreciate the great support and encouragement given by Helen Sargeant who read and corrected the manuscript as well as assisting on several photographic expeditions.

We should also like to extend our thanks to those people who helped us obtain the photographs: Dr. Stephen Smith, Steve Cronin, Nick Mooney, Nigel Brothers, of the Department of Lands, Parks and Wildlife, also to Terry (Scobie) Pye and Graeme Challis for helping to locate particular small mammals. Also to Mike Jagoe of the Tahune Wildlife Park, Barbara Jones of Bonorong Park and Bob Reeves of the East Coast Bird Park for allowing us access to their animals.

Dee Smith and Rosie Bickel gave freely of their time to type the manuscript, Jason Grainger and Janelle Longman greatly assisted.

CONTENTS

Uncaptioned photographs: Front cover, juvenile Tasmanian Devil; Rear cover, Forester Kangaroo; Title page, Little Pygmy-possum.

FOREWORD

By Jessie Luckman

The Tasmanian Conservation Trust is to be commended for producing this handy pocket-sized book on the native mammals of Tasmania. It should prove an informative guide for the growing number of people who are becoming interested in and concerned about our native fauna.

One animal which has received much publicity in recent years is the Thylacine or Tasmanian Tiger. From the earliest days of settlement it was trapped, hunted and shot until it is now on the brink of extinction and has, belatedly, been given the status of "wholly protected".

About half the remaining animals described here are similarly protected — on paper. There are also eight of which can still be said "little is known" of their habitat requirements.

In view of the extensive use of Tasmanian forests for paperpulp and woodchip exports and present day clearfelling methods of forest harvesting, it is significant that twenty-nine of the thirty-eight native mammals feed and breed in forest areas. Seventeen of these make their nests in hollow logs or in hollows of trees and their branches; that is, in mature long established trees. Our mature forests are being cut out at an alarming rate, with the total removal of old trees, understorey shrubs and forest litter. By this destruction of their habitat, leaving them no nesting places, no food, nowhere to go, these animals are condemned to die just as surely as if people were encouraged to shoot them.

Only through well informed public concern and outspoken action will the future survival of these animals be assured.

INTRODUCTION

Tasmania can rightly be called the "Jewel in the Crown" of Australia when referring to its mammals. Many species of mammals, once widely distributed throughout southern Australia, have either disappeared or are sadly reduced to remnant populations.

In Tasmania though, a combination of isolation (which has enabled the exclusion of such devastating introduced predators as the European Fox) and the rugged, mountainous terrain (which has slowed down the ever intruding process of land clearance) has enabled the great majority of native mammals to survive in healthy populations.

Predators like the Tasmanian Devil, once widespread on mainland Australia although having died out there many centuries ago are thriving and common in Tasmania. Similarly the Barred Bandicoot which is now restricted to one small colony in Victoria, and terribly endangered there, can be regarded as abundant in many parts of Tasmania. The sight of so many dead animals killed by vehicles on Tasmanian roads, whilst distressing, is an indication of the healthy state of the populations of many of our mammals.

The objective of producing this volume is not simply to produce a picture book of Tasmanian mammals, but to help create an awareness of the animal treasures which co-exist with us on this island. With very little effort it is possible to observe and enjoy the private lives of so many native species in their natural habitat.

This book brings together, for the first time, an outstanding series of colour photographs of all the native Tasmanian mammals including the eight species of bats which are so often overlooked. The majority of photographs were taken especially for this book by Dave Watts. This involved many weeks of travel to the four corners of Tasmania and many hours of patiently waiting in the darkness. The results speak for themselves. Photographs taken by photographers other than Dave are credited where they appear.

This is essentially a guide to Tasmania's native land mammals and we have therefore not included the following introduced mammals; Fallow Deer, *Dama dama;* Hare, *Lepus capensis;* Rabbit, *Oryctolagus cuniculus* and Feral Cat, *Felis catus.* We have included three introduced small mammals, the Black Rat, *Rattus rattus;* Brown Rat, *Rattus norvegicus* and House Mouse, *Mus musculus;* as these can easily be confused with native rats and mice.

It is also outside the scope of this book to include details of the marine mammals inhabiting the coastal waters of Tasmania. The only common resident seal is the Australian Fur Seal, *Arctocephalus pusillus.* This fur seal may be encountered singly or in groups on rocky islands.

Also occasionally seen on Tasmanian beaches as vagrants are the Southern Elephant Seal, *Mirounga leonina;*and the Leopard Seal, *Hydrurga leptonyx.* At least twenty-eight species of cetaceaus whales and dolphins live in or migrate through Tasmanian waters.

Tasmanian mammals are divided into three major groups, Monotremes, Marsupials and Eutherian or Placental mammals. Monotremes (Platypus and Echidna) differ from other mammals in that the females lay soft-shelled eggs and milk is supplied to the young after they hatch by means of numerous ducts opening into the skin of the abdomen, not via nipples. The name Mono (one) treme (hole) refers to the single body opening through which pass both faeces and urine. The female also lays her eggs through this opening but the male does have a separate penis.

Marsupials give birth to their young when they are very small and not fully developed. These tiny young crawl unaided through the fur to cling to a nipple, normally protected within a pouch, until fully formed. The name marsupial comes from a latin word meaning pouch. Most native Tasmanian mammals belong to this group.

Eutherian mammals include humans and most domestic animals. Young are carried inside the mother where they are fed and nourished by means of a placenta. They are born fully formed but require further feeding via the mother's nipples until they are large enough to become independent. The active rodents and the bats belong to this group.

All lengths and weights that have been used in this book are the average lengths and weights of mature animals except in the case of bats where a range of lengths and weights, and forearm measurements are given, as these measurements are vital for identification. It is the case with most of these mammals that the male is larger and heavier than the female.

Most Tasmanian native mammals are wholly protected under the National Parks and Wildlife Act (1970). Permits may be issued by the Department of Lands, Parks and Wildlife for the trapping of animals for study purposes and also for the culling of some species which may be so numerous that they compete with humans or cause damage.

Our greatest concern must be about the continuing alteration and destruction of wildlife habitat outside the National Parks, through clearance for agriculture, drainage of wetlands, subdivision and the uncontrolled spread of urban sprawl. Certain forestry operations such as clearfelling and the woodchipping industry can have profound effects on wildlife by removing those old trees which contain hollows vital for many small marsupials and bats in particular.

The text of this guide has been designed to complement the photographs and will provide the reader with clear, easily-read information on identification, habits, habitat, distribution and status of the many mammals. This book is intended to be used as a reference quide to be kept in the car or carried in a rucksack. We hope that a greater awareness, interest and respect for our natural environment will ensure that the great variety of mammals to be seen on these pages will survive to be admired by future generations.

Bob Burton

Dry sclerophyll forest

HABITATS

Rob Blakers

Cool temperate rainforest

Sedgeland

Ted Mead

Coastal heathland

MONOTREMES

Platypus

Ornithorhynchus anatinus

Order MONOTREMATA Family Ornithorhynchidae

Size	Head and Body	430 mm
	Tail	120 mm
	Weight	2.5 kg

Description: The Platypus can be distinguished from other swimming mammals (e.g. Water Rat, see page 78) by its smooth action, low silhouette and absence of visible ears. The broad tail, webbed feet and duck-like bill of the Platypus are easily recognised when out of water. Its soft, dense fur is grey to brown on the back and creamy on the underside. The male has a venomous spur on each ankle and is capable of painfully wounding humans.

Habits: Most active in early morning and late afternoon. Habitat Fresh-water lake or river systems where it has a short residential burrow, usually hidden under a tangle of tree roots just above the water line.

Food: Worms, insects, molluscs and small vertebrates. Food collected on the bottom is stored in large cheek pouches until the Platypus comes to the surface and consumes its catch.

Breeding: In October and November prior to egg laying, the female constructs an elaborate nesting burrow up to 20 m in length. Usually two eggs are laid which hatch one to two weeks later. The young are fed on milk secreted through numerous ducts opening into the mother's abdomen. The young are weaned at four to five months.

Distribution: Tasmania, south east of South Australia, Victoria, eastern New South Wales, north and south eastern Queensland.

Status: Wholly protected by law. However dams, riverbank alteration, pollution and many other alterations to rivers reduce suitable habitat. Common.

Echidna

Tachyglossus aculeatus

Order MONOTREMATA Family Tachyglossidae

Size	Length	400 mm
	Weight	4 kg

Description: The Echidna has a long tubular snout, strong claws and reddish-brown to black fur. Protruding through the fur on the back are stout spines. Males have a venomous spur on each hind leg. The Tasmanian Echidna has more fur and less spines than the mainland form.

Habits: The Echidna usually prefers to eat in the early morning and late afternoon, though in cooler months it can be active during the day. When disturbed it may curl into a ball protected by its spines, or dig into the soil leaving only its spines exposed.

Habitat: Throughout Tasmania, but most often seen in drier areas.

Food: Ants, termites and other small invertebrates which it traps on its long tongue with sticky saliva.

Breeding: Late winter to spring. A soft-shelled egg is laid. The young hatches after about ten days, and is weaned after three months. The female may occupy a burrow while suckling her young, although generally Echidnas have no definite nest site.

Distribution: Tasmania and all mainland states.

Status: Wholly protected by law. The species is secure. Common.

MARSUPIALS

Spotted-tailed Quoll

(Tiger Cat)

Dasyurus maculatus

Order POLYPROTODONTA   Family Dasyuridae

Size	Head and body	600 mm
	Tail	500 mm
	Weight	4 kg

Description: Thick, coarse, golden brown to dark brown fur on the back with a lighter sandy coloured fur underneath. Body and tail covered in spots of different sizes. This is the main way of distinguishing this quoll from the Eastern Quoll (page 22), the latter having no spots on its tail.

Habits: Nocturnal. Although an agile climber it spends most of its time on the forest floor. It is slower-moving and has a quieter temperament than the Eastern Quoll.

Habitat: Favours the wetter forest areas of the Western Highlands, the north and north east.

Food: Small birds, mammals, invertebrates and reptiles. Also scavenges carrion.

Breeding: Late winter or early spring. Litters of around five young are born and remain in the pouch for about two weeks. Young are weaned by eighteen weeks.

Distribution: Tasmania, Victoria, New South Wales and Queensland.

Status: Wholly protected. However land clearance and direct competition with the feral cat may be a threat to this species. Common in suitable habitat.

Eastern Quoll

(Native Cat)

Dasyurus viverrinus

Order POLYPROTODONTA   Family Dasyuridae

Size		
Size	Head and Body	350 mm
	Tail	230 mm
	Weight	1300 g

Description: The Eastern Quoll has thick, dark fawn to black fur, spotted with white. Unlike the Spotted-tailed Quoll it has no spots on its tail which is more rufus in colour and has a whitish tip. The Eastern Quoll is considerably smaller than the Spotted-tailed Quoll and has comparatively larger ears and eyes.

Habits: Mainly nocturnal and less arboreal than the Spotted-tailed Quoll. It mostly hunts on the ground for its food.

Habitat: Dry sclerophyll forest, shrub and heathland. Also common in agricultural areas.

Food: Mainly invertebrates particularly pasture pests, eg cockchafer and corbie grubs. Also small mammals and birds. Grasses and fruit.

Breeding: Mates between mid May to early June. Typical litter size is six. Young remain in pouch for eight to ten weeks. After this they are left in grass-lined den in burrows, hollow logs or rock piles while the mother forages. The young are weaned by the end of October.

Distribution: Widespread in Tasmania, but possibly no longer exists on the mainland.

Status: Wholly protected. Has successfully adapted to European settlement in Tasmania. Common.

Tasmanian Devil

Sarcophilus harrisii

Order POLYPROTODONTA Family Dasyuridae

Size	Head and body	600 mm
	Tail	250 mm
	Weight	8 kg

Description: Coarse, thick black fur usually with irregular white markings on the neck, shoulders and rump areas. They have very powerful jaws giving them a fierce appearance.

Habits: The Devil is generally nocturnal. It runs with an awkward lope and leaves an easily recognisable track of one print, then two side by side, then one print (see page 97). They can frequently be heard growling and screaming when squabbling with others over a carcass. Young devils are adept at climbing trees. During the day Devils shelter in caves, old burrows and thick scrub.

Habitat: Forest, woodland and agricultural areas.

Food: Mainly a carrion eater, scavenging anything of animal origin, but probably also takes live prey. Because of its powerful teeth and jaws, it can consume most of a carcass including the bones.

Breeding: Mates in March and gives birth in April. Up to four young per litter. The young remain in the pouch for fifteen weeks and are completely weaned at about thirty weeks.

Distribution: Tasmania. Once widespread on the mainland but died out many centuries ago.

Status: Wholly protected. Common. Abundant in the north east.

Dusky Antechinus

(Dusky Marsupial Mouse)

Antechinus swainsonii

Order POLYPROTODONTA Family Dasyuridae

Size		
Size	Head and Body	120 mm
	Tail	100 mm
	Weight	65 g

Description: Soft and dense fur, deep brown-black on top and greyish white below. It has small eyes, small ears and long claws, and its muzzle is long and narrow.

Habits: Forages both by day and night. Moves in a jerky fashion and may be heard uttering a "siss" cry or a "chit" sound.

Habitat: Rainforest and wet sclerophyll, particularly where there is dense leaf litter and an abundance of ferns.

Food: Soil invertebrates, fruit, lizards and carrion. Locates its prey principally by smell.

Breeding: Females construct grass-lined nests in creek banks or under logs and decaying vegetation. Mating occurs in early spring and all males die within three weeks of mating. The usual litter size is eight, with the young remaining in the pouch for up to eight weeks. The young are independent by about three months of age.

Distribution: Tasmania, Victoria and New South Wales.

Status: Wholly protected. Common within its forest habitat.

Swamp Antechinus

(Little Marsupial Mouse)

Antechinus minimus

Order POLYPROTODONTA Family Dasyuridae

Size		
	Head and body	120 mm
	Tail	80 mm
	Weight	55 g

Description: Coarse and grizzled fur which has a grey/reddish tinge on top and greyish yellow underneath. It has short ears, small eyes and long foreclaws. Its tail is short and covered with short hair.

Habits: Uses its long foreclaws to dig for food. Forages by day and night.

Habitat: Wet button-grass sedgeland, but also found in coastal heathland.

Food: Insects and larvae, lizards, earthworms and spiders.

Breeding: September to December. Normal litter size is six. As with the Dusky Antechinus, males die soon after their first mating.

Distribution: Tasmania, King and Flinders Island, south east of South Australia and southern coastal areas of Victoria.

Status: Wholly protected. Threatened by destruction of its limited, preferred habitat, particularly by too frequent burning. Common.

White-footed Dunnart

Sminthopsis leucopus

Order POLYPROTODONTA Family Dasyuridae

Size		
	Head and body	100 mm
	Tail length	70 mm
	Weight	28 g

Description: Dark brown fur above and light grey to white below. It has large round ears, a slender tail and white paws.

Habits: It is very quick-moving and is mostly nocturnal.

Habitat: Widespread, but with patchy distribution in a range of habitats including rainforest, sclerophyll forest and heathland.

Food: Invertebrates, small reptiles and possibly small mammals.

Breeding: Nests appear to be in bark-lined hollows in trees and logs. Young are born in late spring and the maximum litter size is eight.

Distribution: Tasmania, Flinders Island, western Victoria, south eastern New South Wales.

Status: Wholly protected. Relatively little is known about the White-footed Dunnart because of the difficulty of capturing it. Probably common.

Thylacine

(Tasmanian Tiger)

Thylacinus cynocephalus

Order POLYPROTODONTA Family Thylacinidae

Size		
Size	Head and body	1200 mm
	Tail	550 mm
	Weight	20 kg

Description: The coarse fur is sandy brown in colour with dark brown stripes extending from the mid-back to the base of the tail. It has a large head with a long muzzle and has a very wide gape. Its tail is semi-rigid. Resembles a medium sized dog.

Habits: Mainly nocturnal but may bask in the sun in cold weather. Moves with a stiff gait and emits a coughing yap when disturbed. Very secretive.

Habitat: Once widespread, favouring sclerophyll forest and heathland. Present distribution is open to question.

Food: Kangaroos, wallabies and other game.

Breeding: In winter or spring. A litter of two or three young is carried in a backward-opening pouch.

Distribution: Tasmania. Once widespread on the Australian mainland, but died out many centuries ago.

Status: Wholly protected. The last documented Thylacine died in captivity in 1936. Since then many sightings have been reported and some authorities consider the Thylacine still exists. The decline of the Thylacine may be due to a disease which also afflicted the Tasmanian Devil, and the bounty paid for every scalp the hunters produced. Very rare.

Photograph courtesy Archive Office of Tasmania

Southern Brown Bandicoot

Isoodon obesulus

Order POLYPROTODONTA   Family Peramelidae

Size		
	Head and body	350 mm
	Tail length	120 mm
	Weight	1.2 kg

Description: Dark, grizzled brown fur above, creamy/white below. It has round ears, long nose and a short pointed tail.

Habits: Nocturnal. Prefers to stay close to cover.

Habitat: Widespread, but prefers scrub or areas of low ground cover which are burnt from time to time.

Food: Soil invertebrates, particularly earthworms and insects and their larvae.

Breeding: Constructs a well concealed nest of grass and other plant material, sometimes mixed with earth. Breeds between June and February. Litter size varies from one to four young which are raised in a rear-opening pouch. At least two litters are produced each season.

Distribution: Tasmania, Victoria, south eastern New South Wales, Northern Queensland, south eastern South Australia and south of Western Australia.

Status: Wholly protected, while land clearance and the reduction in the frequency of small-scale fires lessens the amount of suitable habitat. It is widely distributed in Tasmania, including agricultural regions. Common in suitable habitat.

Eastern Barred Bandicoot

Perameles gunnii

Order POLYPROTODONTA Family Peramelidae

Size		
	Head and body	320 mm
	Tail	80 mm
	Weight	950 g

Description: The fur of the Eastern Barred Bandicoot is soft and dense. It is grey on top with paler bands across the rump and is a lighter grey underneath. It has big ears, a white tail and long claws on its forefeet.

Habits: Mostly nocturnal. It moves at a rapid gallop or bound, and can clear over one metre at a leap.

Food: Soil invertebrates, including pasture pests e.g. cockchafer larvae and corbie grubs. It also eats berries in season.

Breeding: Late May to March. Litter size is one to four, raised in a rear-opening pouch. Young animals are independent three and five months after birth.

Distribution: Tasmania, south western Victoria and Furneaux Group.

Status: Wholly protected. It has adapted well to European settlement in Tasmania but has almost died out on the Australian mainland. Common.

Common Wombat

Vombatus ursinus

Order DIPROTODONTA Family Vombatidae

Size		
	Head and body	1000 mm
	Tail	25 mm
	Weight	26 kg

Description: Coarse, dark brown fur. Solid and powerfully built body. Short legs with broad paws and strong claws. Short ears, small eyes and a large naked nose.

Habits: Largely nocturnal but in cold weather may be seen grazing or basking in the sun. Dung is easily recognizable, being rectangular and usually deposited on rocks or logs.

Food: Native grasses, herbs, shrubs and succulent roots.

Breeding: Constructs elaborate burrows up to 20m long with interconnecting tunnels. Breeding occurs at any time of the year but usually in winter. Generally one young is born, remaining in the pouch for six months, then staying with the mother until it is about eighteen months old.

Distribution: Tasmania, south eastern South Australia, Victoria and New South Wales.

Status: Partly protected. The species is secure at present, although suffering a reduction of range through clearing for agriculture. Common in suitable habitat.

Common Ringtail Possum

Pseudocheirus peregrinus

Order DIPROTODONTA   Family Petauridae

Size		
	Head and body	330 mm
	Tail	330 mm
	Weight	950 g

Description: Reddish brown fur above; lighter below. Short ears with a white patch behind. Distinguished from the Brushtail Possum (page 44), by its tapering, white-tipped tail.

Habits: Nocturnal. A vocal animal with a soft, high-pitched twittering call.

Habitat: Widespread throughout Tasmania. Most abundant in areas of tall dense tea-tree.

Food: Leaves, flowers (particularly eucalypts) and fruits.

Breeding: Builds a spherical, bark or grass-lined nest in hollows, forks of trees and amongst branches. Breeding occurs from April to November with two young usually born. These remain in the pouch for four months and are fully weaned at six months. Once out of the pouch, the young may be carried on the mother's back while she forages.

Distribution: Tasmania, south eastern South Australia, Victoria, New South Wales, Queensland and south west of Western Australia.

Status: Wholly protected. Numbers declined during the 1950's but are now recovering. Common.

Sugar Glider

Petaurus breviceps

Order DIPROTODONTA Family Petauridae

Size		
	Head and body	170 mm
	Tail	190 mm
	Weight	120 g

Description: Soft, dense, blue-grey coloured fur with a dark stripe down the centre of its back. Bushy tail sometimes tipped with white. A flap of loose skin extends along the side of the body from front paw to back paw which allows the animal to glide for a distance of up to 50m from the tallest trees.

Habits: Nocturnal. The Sugar Glider has a range of calls, a shrill yapping to warn of danger, a scream when fighting and a defiant chatter when disturbed. In very cold weather it may conserve energy by huddling with others in a nest hollow or by becoming torpid.

Habitat: Open forest and woodland areas. Lives in groups with up to seven adults and their young sharing a nest.

Food: Invertebrates, flowers, nectar and gum produced by acacias.

Breeding: Constructs a leaf-lined nest in hollows of old trees. Breeding begins in August. Usual litter size is two and these remain in the pouch for two to three months, then spend another month in the group nest.

Distribution: Tasmania and all mainland states. Absent from much of the west and central mainland. Probably introduced into Tasmania.

Status: Wholly protected. Species seems secure at present, but may undergo a reduction in numbers with clearfelling and removal of old nesting trees. Common.

Common Brushtail Possum

Trichosurus vulpecula

Order DIPROTODONTA   Family Phalangeridae

Size		
Size	Head and body	450 mm
	Tail	300 mm
	Weight	3.5 kg

Description: The Brushtail Possum has very thick soft fur ranging in colour from silver-grey to black, to reddish and sometimes golden. It has long oval ears and a bushy tail with a short, naked area under the tip which distinguishes it from the Ringtail Possum (Page 40) which has a tapering white-tipped tail.

Habits: Nocturnal. Utters sharp hisses or deep guttural coughs, particularly in the breeding season. Feeds on the ground and is less arboreal in Tasmania compared with mainland Australia.

Habitat: Most bush areas and in agricultural and suburban areas.

Food: In dry forests — grasses, herbs, acacias and eucalyptus leaves: In wet forest — ferns, grasses and myrtle leaves.

Breeding: Between March and June with a secondary breeding period between August and October. A single young is born, and remains in the pouch for up to six months. A further two months are spent riding on the mother's back prior to breeding.

Distribution: Tasmania, and all mainland states.

Status: Partly protected species which has generally benefited from current land management practices. Abundant in most areas.

Eastern Pygmy-possum

Cercartetus nanus

Order DIPROTODONTA   Family Burramyidae

Size		
	Head and body	90 mm
	Tail	90 mm
	Weight	30 g

Description: Fur is pale fawn-grey above and lighter coloured below. Prehensile tail which is naked for most of the length. It has oval shaped ears and pinkish feet and nose.

Habits: Nocturnal. It is an agile climber. In winter when food is short much of its time is spent in a state of torpor. Torpor may last for two weeks at a time.

Habitat: Forest areas. From rainforest to sclerophyll forest.

Food: Insects, spiders, small lizards; also nectar and pollen from banksias, bottlebrushes and eucalypts gathered with a brush tipped tongue.

Breeding: Nests lined with shredded bark are constructed in very small tree holes or in tree forks.

Distribution: Tasmania, Victoria, eastern New South Wales and south eastern Queensland.

Status: Wholly protected. Threatened by clearfelling of old-growth mature forest which contains nesting hollows.

Little Pygmy-possum

Cercartetus lepidus

Order DIPROTODONTA Family Burramyidae

Size	Head and body	60 mm
	Tail	70 mm
	Weight	7 g

Description: Soft pale fawn coloured fur above, light grey below. Large ears and eyes and a prehensile tail.

Habits: Nocturnal. Like the Eastern Pygmy-possum, it undergoes periods of torpidity, particularly in winter when food is scarce. Spends most of its time close to the ground or in dense scrub to avoid attacks from owls.

Habitat: All forest types except rainforest; favours dry sclerophyll forest.

Food: Insects, spiders and small lizards; probably also nectar, blossoms and fruit.

Breeding: Builds bark lined nests, usually in small hollows in trees. Young are born between September and January with up to four per litter. After about six weeks the young are too big for the pouch and may either be left in the nest while the mother forages, or travel clinging to her fur. Young are mature at about three months.

Distribution: Tasmania, Victoria, South Australia, Kangaroo and Flinders Islands.

Status: Wholly protected. Extensive clearing of sclerophyll forests lessens the available habitat. Forests regenerated after clearfelling are unlikely to provide suitable nesting sites.

Long-nosed Potoroo

Potorous tridactylus

Order DIPROTODONTA Family Potoroidae

Size	Head and body	360 mm
	Tail	230 mm
	Weight	1.3 kg

Description: Red-brown to grey coloured fur above and paler below. Long tapering nose with bare patch of skin above the nostrils. Short rounded ears. Some have a white tip to their short tail.

Habits: Mostly nocturnal but may be seen at dusk. When moving quickly it hops on its hind legs.

Habitat: Widespread in sclerophyll forest and heathland which provides thick ground cover through which the Potoroo forms a network of runways. Prefers areas with light sandy soil.

Food: Roots, tubers, fungi, insects and soil invertebrates.

Breeding: Has two breeding seasons — in winter/early spring and late summer. A single young is produced at a time and remains in the pouch for four months.

Distribution: Tasmania, Victoria, New South Wales, King and Flinders Islands.

Status: Wholly protected. Much habitat has been reduced by land clearing. Common in areas of remaining habitat.

Tasmanian Bettong

Bettongia gaimardi

Order DIPROTODONTA   Family Potoroidae

Size	Head and body	320 mm
	Tail	320 mm
	Weight	2 kg

Description: Coarse brownish-grey coloured fur above; greyish white underneath. Short round ears and long tail which usually has a white tip.

Habits: Nocturnal. Individuals sometimes travel over a kilometre from the nest to a feeding area.

Habitat: Widespread but patchily distributed in eastern Tasmania in dry sclerophyll forest and woodland with an open understorey.

Food: The bulk of the Bettong's diet is the fruiting bodies of underground fungi. Also eats seeds, mushrooms, insects and gum from acacia shrubs.

Breeding: Builds a woven nest of dry grass and bark in a hollow under a fallen tree or among bushes. Material for the nest construction is carried in the Bettong's tightly-coiled tail.

Breeding: occurs all year round. Females usually produce two to three young per year. The young are usually in the pouch for three to four months and are weaned at six months.

Distribution: Tasmania.

Status: Wholly protected. Regenerated forests are unsuitable for Bettongs and large areas of habitat are being lost. The few remaining high density populations are on private land. Common in some areas.

Tasmanian Pademelon

(Rufous Wallaby)

Thylogale billardierii

Order DIPROTODONTA Family Macropodidae

Size		
Size	Head and body	600 mm
	Tail	410 mm
	Weight	7 kg

Description: The dense fur is dark-brown or grey-brown above and reddish-brown underneath. Males are much larger than females and are considerably more muscular in forearms and chest.

Habits: Nocturnal.

Habitat: Widespread in all areas where there is dense undergrowth, preferring forest adjacent to cleared areas for feeding. Shelters in dense vegetation during the day.

Food: Grasses, herbs and taller shrubs.

Breeding: Breeding is continuous, but most of the young are born in early winter. One young is born and remains in the pouch for six months.

Distribution: Tasmania.

Status: Partly protected. Large numbers apparent in forestry and agricultural areas. This species once occured on the mainland but now appears to have died out. Abundant.

Red-necked Wallaby

(Bennett's Wallaby)

Macropus rufogriseus

Order DIPROTODONTA Family Macropodidae

Size		
Size	Head and body	800 mm
	Tail	750 mm
	Weight	16 kg

Description: Soft thick fur dark grey above and paler below except on the neck where it is reddish-brown. Black nose and paws.

Habits: Mostly nocturnal but can be seen at dusk. Usually solitary, although they may feed close to each other.

Habitat: Widespread throughout Tasmania in open forests, heathland and sedgeland, particularly in areas adjacent to pasture. Rests during the day among dense patches of scrub.

Food: Grasses and herbs.

Breeding: Young are born between January and July with most females giving birth in late summer. Usually one young is born and remains in the pouch for nine months. Weaning is about seventeen months.

Distribution: Tasmania, Victoria, south eastern South Australia, eastern New South Wales, south eastern Queensland, King and Flinders Islands.

Status: Partly protected. The species appears to have benefited from European settlement with large numbers in forestry and agricultural areas. Abundant.

Eastern Grey Kangaroo

(Forester Kangaroo)

Macropus giganteus

Order DIPROTODONTA   Family Macropodidae

Size		
	Head and body	1500 mm
	Tail	940 mm
	Weight	45 kg

Description: The Forester Kangaroo has soft, brownish-grey fur above and paler coloured fur underneath. It has a hairy muzzle with fine hairs between the nostrils and upper lip. Relatively large ears.

Habits: Active in the evenings and at night. May be heard making a series of clucking sounds, or if alarmed a guttural cough.

Habitat: It is the only kangaroo occurring in Tasmania, and is found in dry sclerophyll forests, woodlands and open grassy plains in the north east and the Midlands. The species has been introduced to Maria Island National Park.

Food: Grasses.

Breeding: Mainly in summer with usually one young born. Young stays in pouch for ten months and is weaned at about eighteen months, by which time the mother may have another young in the pouch.

Distribution: Tasmania, Victoria, New South Wales, eastern South Australia and Queensland.

Status: Wholly protected. Populations were greatly depleted by hunting and the extensive agricultural development of the Tasmanian Midlands. Active measures are being taken to ensure the continued survival of the species in Tasmania. Common in a few selected areas.

PLACENTAL MAMMALS

Greater Long-eared Bat

Nyctophilus timoriensis

Order CHIROPTERA Family Vespertilionidae

Size		
	Head and body	65 mm
	Tail	50 mm
	Forearm length	43.5-48 mm
	Weight	9.8-18.9 g

Description: Dark brown fur above; lighter brown below. Ears large and joined over the head.

Habits: Nocturnal. Hibernates from late autumn to early spring.

Habitat: A wide range of forest habitats, where it roosts in trees during the day, often with others of the same species.

Food: Insects, mostly non-flying ones which are collected off vegetation and from the forest floor. It feeds flying slowly close to the ground and amongst the undergrowth.

Breeding: One or two young are born in late spring or early summer and are weaned by mid-summer.

Distribution: Tasmania, Victoria, New South Wales, Queensland, South Australia and Western Australia.

Status: Wholly protected. Forest regenerated after logging operations are unlikely to provide as many suitable roosting sites as mature forest. In the long-term a decline in this species may result. Uncommon. This bat species was previously named N. gouldii.

GB Baker/Australian Nature Transparencies

Lesser Long-eared Bat

Nyctophilus geoffroyi

Order CHIRIPTERA   Family Vespertilionidae

Size		
	Head and body	0-50 mm
	Tail	43 mm
	Forearm	36.5-42.6 mm
	Weight	6.8-12.2 g

Description: Light grey-brown fur above and pale below. Ears are very long and are joined above the forehead. May be distinguished from the Greater Long-nosed Bat by its smaller size, paler fur and by its well developed Y-shaped nose leaf.

Habits: Nocturnal. Hibernates from late autumn to early spring.

Habitats: Widespread in all forest types, including the South — West. It roosts in trees, often with others of the same species. This species has also been recorded foraging and roosting in towns.

Food: Insects, mostly flying ones. It feeds by flying slowly close to the ground, sometimes alighting in the undergrowth.

Breeding: Young are born late spring or early summer, with a usual litter size of two. Young are weaned by mid-summer.

Status: Wholly protected. Forests regenerated after clearfelling operations are unlikely to provide as many suitable roost sites as mature forests, therefore a long-term decline in the species is likely. Common.

Gould's Wattled Bat

Chalinolobus gouldii

Order CHIRIPTERA Family Vespertilionidae

Size		
	Head and body	65-75(70) mm
	Tail	40-50(45) mm
	Forearm	44.0-48.7 mm
	Weight	12.2-17.8 g

Description: Fur on the dorsal surface is dark brown tending to black on the head and shoulders; slightly paler fur underneath. Fleshy lobes at the base of the ear and corner of the mouth. It has an elevated forehead.

Habits: Nocturnal. Roosts in hollow trees, usually in colonies. Hibernates from late autumn to early spring.

Habitat: A wide range of forest types, although not recorded in South-West Tasmania.

Food: Insects. Feeds by flying quickly either in upper layers of forest canopy or above it.

Breeding: Young are born in late spring or early summer with litter size usually two. The young attach themselves to the nipples and remain attached while the mother is flying. Weaning is completed by early February.

Distribution: Tasmanian and all mainland states.

Status: Wholly protected. Forests regenerated after logging operations are unlikely to provide as many suitable roosting sites as mature forests. In the long-term a decline in the species may result. Uncommon.

Bruce Thomson/Australian Nature Transparencies

Chocolate Wattled Bat

Chalinolobus morio

Order CHIROPTERA   Family Vespertilionidae

Size		
	Head and body	50-60(57) mm
	Tail	45-50(48) mm
	Forearm	36.2-43 mm
	Weight	6.7-13.7 g

Description: Chocolate brown coloured fur all over. Ear lobes not as conspicuous as that of the Gould's Wattled Bat.

Habits: Nocturnal. Hibernates in the winter but its period of hibernation is shorter than the other species of Tasmanian bats. Usually roosts in trees, often communally.

Habitat: All forest types including rainforest.

Food: Insects collected mostly in the middle layers of the forest between the canopy and understorey.

Breeding: Young are born in late spring and early summer with usually one young per litter. Weaning is complete in February, usually later than other species.

Distribution: Tasmania and all mainland states.

Status: Wholly protected. However, forest regenerated after logging operations are unlikely to provide as may suitable roosting sites as mature forests. This may result in a long term decline of the species. Common.

Tasmanian Pipistrelle

Falsistrellus tasmaniensis

Order CHIROPTERA Family Vespertilionidae

Size		
Size	Head and body	65 mm
	Tail	40 mm
	Forearm length	46-52.2 mm
	Weight	16.2-24.1 g

Description: Reddish-brown fur above and a paler brown on the belly. Ears are long and pointed with a notch at the top. It is the largest bat found in Tasmania.

Habits: Nocturnal. Hibernates from late autumn to early spring. Roosts in trees, often communally. Flies fast and direct.

Habitat: Most forest types, but has also been found foraging and roosting in urban environments.

Food: Insects; mainly beetles. Feeds in upper layers of forest canopy or above it.

Breeding: A single young is born in late spring or early summer. Weaning is completed by early February.

Distribution: Tasmania, Victoria, New South Wales, south eastern Queensland and south west of Western Australia.

Status: Wholly protected. However forests regenerated after logging operations are unlikely to provide as many suitable roosting sites as mature forests. This may result in a long-term decline of this species. Uncommon.

GA & MM Hoye/NPIAW

King River Eptesicus

Eptesicus regulus

Order CHIROPTERA   Family Vespertilionidae

Size		
	Head and body	45-55(47) mm
	Tail	34-42(38) mm
	Forearm	31.0-35.6 mm
	Weight	3.8-7.7 g

Description: Reddish brown dorsal fur with lighter brown on the ventral surface. Older animals sometimes lack the reddish tinge to their fur. The skin over the forearm is the same colour as that on the wing.

Habits: Nocturnal. Roost in trees, often communally.

Habitat: Wide range of forest habitats.

Food: Insects collected from the top of the forest understorey.

Breeding: One young born in the late spring or early summer. Weaning is completed by mid February.

Distribution: Tasmanian, Victoria, New South Wales, south east of South Australia and south of Western Australia.

Status: Wholly protected. Forests regenerated after logging operations are unlikely to provide as many suitable roosting sites as mature forests and this may lead to the long-term reduction of all Eptesicus species. Common.

Little Forest Eptesicus

Eptesicus vulturnus

Order CHIROPTERA Family Vespertilionidae

Size		
Size	Head and body	40-50(44) mm
	Tail	30-40(34) mm
	Forearm	27.2-31.7 mm
	Weight	3.7-5.6 g

Description: The smallest Tasmanian bat. Dorsal fur is mid-dark grey and ventral fur is grey with distinct pale tips. The tragus is translucent white. The skin over the forearm is distinctly lighter than on the wing membrane.

Habits: Nocturnal. Roosts in trees often communally.

Habitat: Wide range of forest habitats but appears to favour lowland situations.

Food: Insects collected from the top of the forest understorey.

Breeding: One young born in late spring or early summer. Weaning is completed by mid February.

Distribution: Tasmania, Victoria, New South Wales, South Australia, Western Australia and Northern Territory.

Status: Wholly protected. Forests regenerated after logging operations are unlikely to provide as many suitable roosting sites as mature forests and this may lead to the long-term reduction of all Eptesicus species. Common.

Large Forest Eptesicus

Eptesicus sagittula

Order CHIROPTERA Family Vespertilionidae

Size		
	Head and body	40-60(50) mm
	Tail	40 mm
	Forearm	32.0-37.0 mm
	Weight	4.8-7.8 g

Description: The following two species of bats are very similar to the Large Forest Eptesicus. The main distinguishing features being the forearm length and the skin colour of the forearm and wing membrane. The Large Forest Eptesicus has dark brown or almost black fur on its dorsal surface and is paler underneath. The skin over the forearm is the same colour as the wing membrane.

Habits: Nocturnal. Roosts in trees often communally.

Habitat: Wide range of forest habitats including rainforest.

Food: Insects collected from the middle layer of the forest between the canopy and understorey.

Breeding: One young born in late spring or early summer. Weaning is completed by mid-February.

Status: Wholly protected. Forests regenerated after logging operations are unlikely to provide as many suitable roosting sites as mature forests and this may lead to the long-term reduction of all Eptesicus species. Common in available habitat.

GB Baker/Australian Nature Transparencies

Water Rat

Hydromys chrysogaster

Order RODENTIA   Family Muridae

Size		
	Head and body	260 mm
	Tail	270 mm
	Weight	600 g

Description: The Water Rat has thick, glossy, water-repellent fur which is dark brown above and golden orange below. Thick tail which usually has a white tip, small eyes, ears and nostrils which are set high on the head. Broad, partly webbed hind feet.

Habits: Nocturnal and crepuscular. Constructs tunnels in banks.

Habitat: Near permanent bodies of fresh or brackish water. Also estuaries and offshore island groups in salt water.

Food: Aquatic insects, fish, crustaceans. Also small mammals and water birds.

Breeding: Throughout the year, but mostly in spring and summer. Up to five litters may be produced annually, each litter has three to four young, which are suckled for one month and then remain with the mother for another month.

Distribution: Tasmania and all mainland states.

Status: Partly protected. Species considered secure. Common.

Long-tailed Mouse

Pseudomys higginsi

Order RODENTIA Family Muridae

Size`	Head and body	130 mm
	Tail	160 mm
	Weight	70 g

Description: Soft dark grey fur above and paler underneath. Long tail is grey above and white below.

Habits: Mostly nocturnal, but may be seen in daylight, particularly in winter. Makes tunnels and runways beneath the forest debris.

Habitat: Most abundant in high rainfall areas of western Tasmania. Also found in wetter forests in eastern Tasmania and beneath screes in sub-alpine areas.

Food: Fungi, insects, spiders, seeds and fruit.

Breeding: Nests are built in decaying logs or stumps. Breeding occurs from late spring to late summer, and two litters, each of three to four young, may be reared in a good season. Young remain with mother until they reach maturity at age of three months.

Distribution: Endemic to Tasmania.

Status: Wholly protected. Appears secure in rainforest habitat and also able to recolonize clearfelled areas. Common in limited habitat.

New Holland Mouse

Pseudomys novaehollandiae

Order RODENTIA Family Muridae

Size		
Size	Head and body	70 mm
	Tail	80 mm
	Weight	25 g

Description: Short thick grey-brown fur above with grey-white fur below. Tail brown above and white below. The New Holland Mouse may be distinguished from the introduced House Mouse (page 94) by its heavier build, its relatively larger eyes and ears, its more convex nose and hunched position when crouching. The lack of a notch in the upper incisor distinguishes this animal positively from a House Mouse.

Habits: Nocturnal.

Habitat: Dry heathlands and open forest in the north east of the state on sandy soils.

Food: Seeds, insects, flowers, leaves and fungi.

Breeding: Nests in deep burrows. Breeding occurs from late winter to early spring. Litter size varies from two to six which are usually mature at seven weeks.

Distribution: North eastern Tasmania and Flinders Island, Victoria and New South Wales.

Status: Wholly protected. However the species is vulnerable to clearing of its restricted heathland habitat and also to any change in the frequency of fire burning the heathland. Common in available habitat, especially after fires.

Broad-toothed Rat

(mouse)

Mastacomys fuscus

Order RODENTIA Family Muridae

Size		
	Head and body	160 mm
	Tail	110 mm
	Weight	120 g

Description: Sandy to dark brown coloured fur above and paler below. General plump appearance. Broad round head and cheek pouches; small round ears and wide molar teeth that give it its common name.

Habits: Mostly nocturnal but can be seen during the day. Constructs runways under dense vegetation.

Habitat: High rainfall areas, including scrub, sedgeland and sub-alpine heathlands.

Food: Sedges, grasses, seeds. Also leaves and bark from shrubs.

Breeding: Well built nests are constructed under logs or in dense undergrowth. Breeds from early spring to mid-summer. Litter size varies from one to three young which are weaned at about five weeks. Females probably bear two litters per season.

Distribution: Tasmania, Victoria and south eastern New South Wales.

Status: Wholly protected. This species depends on areas of wet scrub sedgeland etc. which have been protected from fire for long periods. Over-frequent burning has led to the loss of much of its habitat. This species is regarded as vulnerable. Uncommon.

Swamp Rat

Rattus lutreolus

Order RODENTIA Family Muridae

Size		
	Head and body	160 mm
	Tail	110 mm
	Weight	120 g

Description: The Swamp Rat has soft thick dark grey-brown fur above and paler coloured fur underneath. Tail is dark grey and scaly. Small ears. May be distinguished from introduced rats (the Black Rat — page 90 and Brown Rat — page 92) by the fact that its tail is much shorter than its head and body and appears plumper.

Habits: Mainly nocturnal, but frequently seen during the day. Constructs tunnels through dense vegetation.

Habitat: Widespread across a range of habitats, including sedgeland, heathland and forest.

Food: Grasses, sedges, ferns, fungi and insects.

Breeding: Nests are constructed in burrows or grass tussocks. Breeds from early spring to autumn with three to five young born per litter. Young are independent after four weeks. Each adult female may have several litters per season.

Distribution: Tasmania, Victoria, eastern New South Wales, south east of South Australia and Queensland.

Status: Wholly protected. The Swamp Rat is probably the most widespread and abundant species of native mammal in Tasmania. Abundant.

INTRODUCED MICE & RATS

Black Rat

Rattus rattus

Order RODENTIA Family Muridae

Size		
Size	Head and body	190 mm
	Tail	230 mm
	Weight	280 g

Description: Brown-black to light brown fur. Sleek elongated body. Large ears, eyes and a pointed snout. Has larger ears, more slender hind-feet and longer tail than the Brown Rat.

Habits: Mainly nocturnal. Good swimmer and climber. Has a fondness for wall cavities and roofs. A timid animal.

Habitat: Widespread, particularly near watercourses, farms, rubbish dumps and warehouses.

Food: Omnivorous.

Breeding: Nests are constructed of any fibrous material. Breeds continuously and females may have six litters of five to ten young each year. Young are weaned at about twenty days.

Distribution: Tasmania and all mainland states.

Status: Introduced. Not protected. Abundant.

IR McCann/Australian Nature Transparencies

Brown Rat

Rattus norvegicus

Order RODENTIA   Family Muridae

Size		
	Head and body	240 mm
	Tail	200 mm
	Weight	320 g

Description: The Brown Rat has grey-brown fur above and is paler coloured underneath. It has a solid build and rather scruffy appearance. Small ears, eyes and a blunt snout.

Habits: Nocturnal. It is known for its aggressive nature and will attack when cornered, unlike the Black Rat and native rats which are timid.

Habitat: Found mainly in coastal and long established urban areas.

Food: Omnivorous.

Breeding: Nests in deep and extensive communal burrows. Breeds continuously throughout the year. Litters number from seven to ten young which are weaned when twenty days old.

Distribution: Tasmania, Victoria southern New South Wales, southern Queensland and south of Western Australia.

Status: Introduced. Not protected. Common.

Greg Fyfe/Australian Nature Transparencies

House Mouse

Mus musculus

Order RODENTIA   Family Muridae

Size		
	Head and body	75 mm
	Tail	80 mm
	Weight	20 g

Description: Brown or yellow-grey fur above and pale grey below. May be distinguished from the native New Holland Mouse by its more concave nose, relatively smaller ears and eyes and more elongated stance. Closer examination will show a notch in the House Mouse's upper incisors. The New Holland Mouse has no such notch.

Habits: Nocturnal.

Habitat: Most common near human habitation, but also widespread in native vegetation, particularly that which has been recently burnt.

Food: Omnivorous but predominantly plants and seeds.

Breeding: Nests are built in burrows or crevices in buildings. Breeds continuously throughout the year. Four to eight young are born per litter. Young are usually sexually mature at eight weeks.

Distribution: Tasmania and all mainland states.

Status: Introduced. Not protected. Abundant.

Expl
Birds
their biolog
WHAT
BIRD IS
THAT?

APPENDIX I

FOOTPRINTS

Information courtesy Department of Lands, Parks & Wildlife.

All are of left feet of adults and are to the same scale

0 1 2 3 4 5 cm

CLIMBERS

Ringtail Possum

Brushtail Possum

Spotted-tailed Quoll

DIGGERS

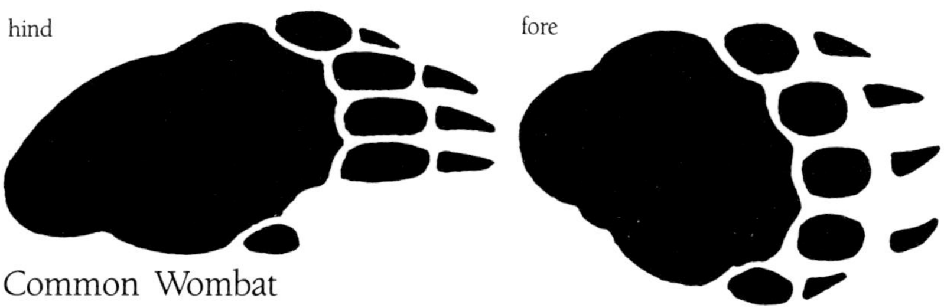

Common Wombat

Echidna

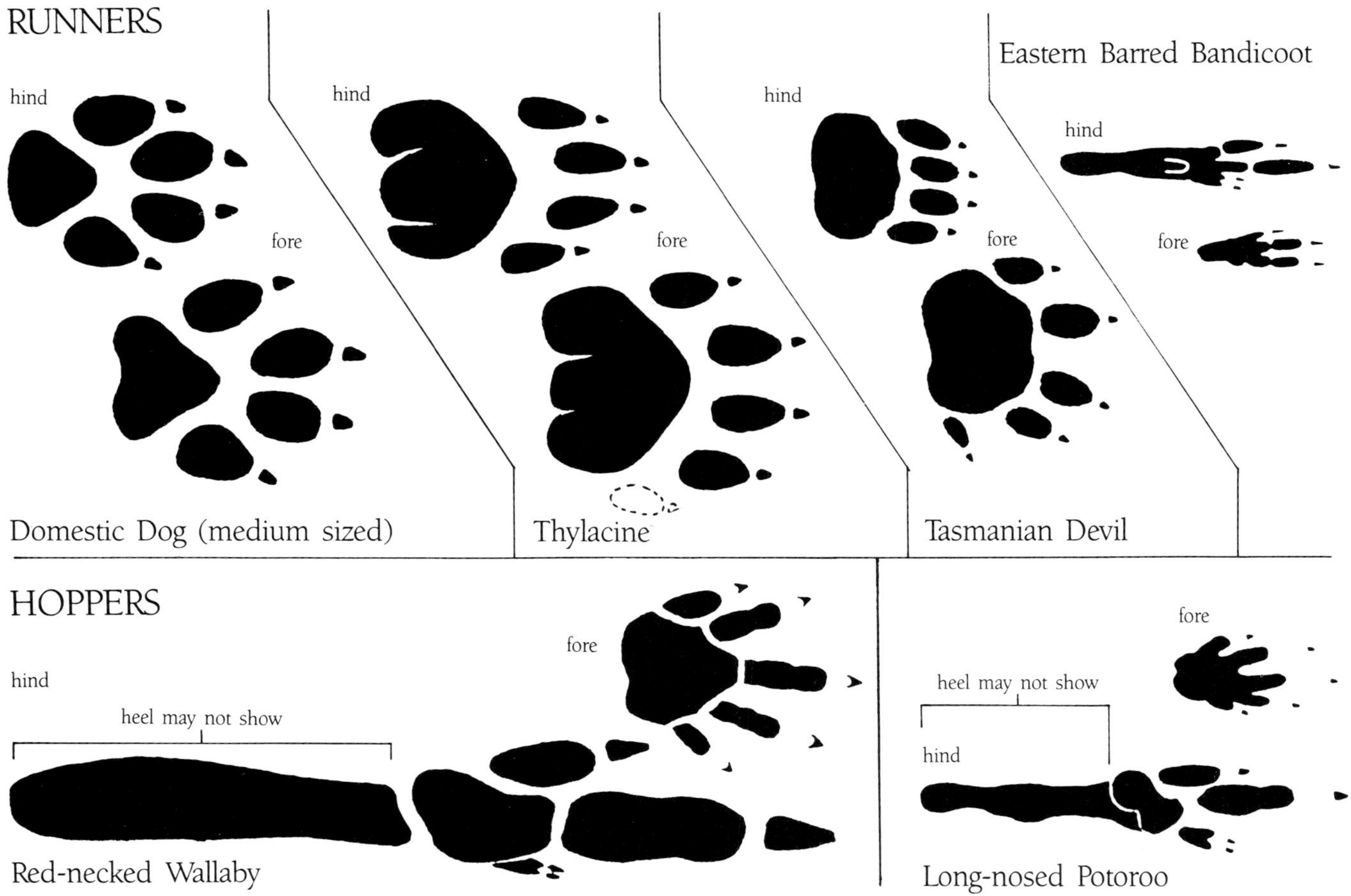
RUNNERS
hind
fore
Domestic Dog (medium sized)
hind
fore
Thylacine
hind
fore
Tasmanian Devil
Eastern Barred Bandicoot
hind
fore
HOPPERS
hind
heel may not show
fore
Red-necked Wallaby
heel may not show
hind
fore
Long-nosed Potoroo

APPENDIX II

BATS

Information courtesy RJ Taylor, MG O'Neil & T Reardon.

Chalinolobus

Eptesicus

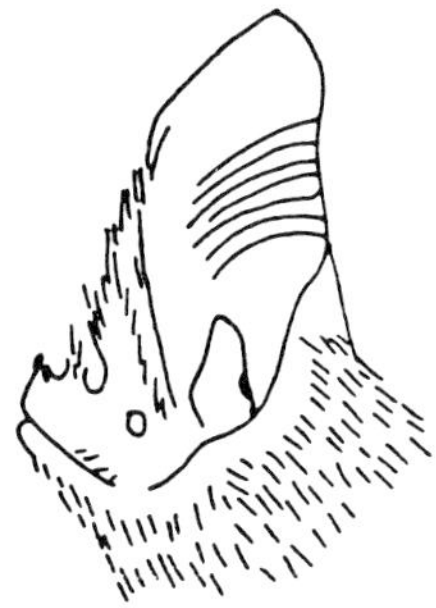

Nyctophilus

Profiles showing the facial features of the four genera of Tasmanian bats.

Falsistrellus

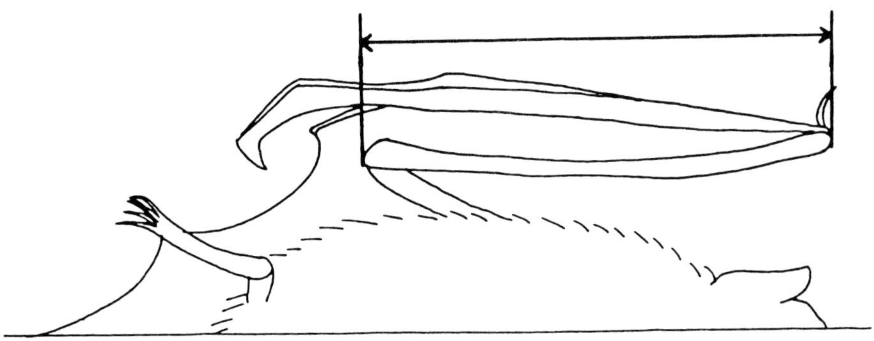

Diagram showing method of measuring forearm length

APPENDIX III

Occurrence of animals within national parks

KEY
★ Present
○ Absent
☆ Unconfirmed

	Asbestos Range	Ben Lomond	Cradle Mt - Lake St Clair	Franklin - lower Gordon	Freycinet	Hartz Mountains	Maria Island	Mount Field	Mount William	Rocky Cape	South-West	Strezlecki	Walls of Jerusalem
Platypus	★	☆	★	★			○	★			★	○	★
Echidna	★	★	★	★	★	★	★	★	★	★	★	★	★
Spotted-tailed Quoll		★	★	★	★	★	○	★	★		★	○	★
Eastern Quoll	★	★	★	★	★	★	○	★	★	★	★	○	★
Tasmanian Devil	★	★	★	★	★	★	○	★	★	★	★	○	★
Dusky Antechinus			★	★	★						★		
Swamp Antechinus		★	★	★	★		★		★		★	★	
White-footed Dunnart	☆				★		☆		★				
Thylacine	○	☆	☆	☆	○	☆	○	☆	☆	○	☆	○	☆
Southern Brown Bandicoot	★			★	★		★		★		★		
Eastern Barred Bandicoot	★		○	○			★				○		○
Common Wombat	★	★	★	★	★	★	★	★	★	★	★	★	★
Common Ringtail Possum	★	★	★	★	★	★	★	★	★	★	★	★	
Sugar Glider	★	★	★		★		○						
Common Brushtail Possum	★	★	★	★	★	★	★	★	★	★	★	★	★

Eastern Pygmy-possum			☆	★			★		★			★	
Little Pygmy-possum	★	★					☆		★		★		
Long nosed Potoroo			★	★	★		★					★	
Tasmanian Bettong	☆	☆	★	○	★	○	★		★	○	○		☆
Tasmanian Pademelon	★	★	★	★	★	★	★	★	★	★	★	★	★
Red-necked Wallaby	★	★	★	★	★	★	★	★	★	★	★	★	★
Eastern Grey Kangaroo	★	★	○	○	○	○	★	○	★	○	○	○	○
Greater Long-eared Bat			★				☆	★					
Lesser Long-eared Bat			★	★	★	☆	☆	★	☆	☆	★	☆	☆
Gould's Wattled Bat							☆						
Chocolate Wattled Bat			★	★	★	☆	☆	★	☆	☆	☆		
Tasmanian Pipistrelle					★		☆	★			★	★	
Little Forest Eptesicus	☆		★	★	★		☆	★	☆	☆	★	☆	☆
King River Eptesicus			★	★	★		☆	★	☆	☆	★	☆	☆
Large Forest Eptesicus	☆		☆	★	★		☆		★	☆	☆		
Water Rat	★				★		★	★	★	★	★		
Long-tailed Mouse		★	★	★		★	☆	★			★	☆	★
New Holland Mouse					★		○					☆	
Broad-toothed Rat			★				○		★		★		
Swamp Rat	★	★	★	★	★	★	★	★	★	★	★	★	★
Black Rat			★		★		★		★			★	
Brown Rat					★		○					☆	
House Mouse	★	★	★	★	★	★	★	★	★	★	★	★	★

GLOSSARY

Arboreal — living in trees

Brackish — slightly salty

Carnivorous — feeding on animal flesh

Carrion — dead and putrifying flesh

Crepuscular — active mainly at dawn and dusk

Dimorphic — occurs in two colour forms

Diurnal — active during the day

Endemic — found in only one region

Invertebrate — an animal without a backbone

Nocturnal — active during the night

Noseleaf — a projecting structure on a bat's nose to direct ultra-sonic sounds uttered through the nostrils

Omnivorous — feeding on both animal and plant matter

Prehensile — being able to grip; normally applied to the tail

Sclerophyll — plants with hard or stiff leaves

Scree — rock debris on the slopes or at base of hills

Sub-alpine — the zone between montane level and tree line

Torpor — breathing slows, body temperature falls and the animal does not respond to normal stimuli

Tragus — a flap situated inside the outer ear

Venomous — poisonous

BIBLIOGRAPHY AND RECOMMENDED READING LIST

THE MAMMALS OF TASMANIA by R H Green (1973), published by the author, Launceston, Tasmania.

AUSTRALIAN MAMMALS: A Field Guide for NSW, VIC, SA, & TAS by J Hyett & N Shaw (1980), published by Thomas Nelson Australia, Melbourne.

THE TASMANIAN TIGER by S Smith (1980), published by Tasmanian NP & WS.

THE COMPLETE BOOK OF AUSTRALIAN MAMMALS by R Strahan (1983) published by Angus & Robertson.

TASMANIAN BATS: Identification, distribution and natural history. By R J Taylor, M G O'Neil & T Reardon.
Papers and Proceedings of the Royal Society of Tasmania, Volume 121 (1987).

MAMMAL TRACKS & SIGNS by B Triggs (1984), published by Oxford University Press, Australia.

RODENTS OF AUSTRALIA by C H S Watts & H J Aslin (1981), published by Angus & Robertson, Australian Natural Science Library.

NOTES

Your personal observations of Tasmanian mammals can help conservation of these species. This book contains the most recent information but there is a lot still to be discovered about our wildlife.

If your observations differ from information given in this book, please send details to either Tasmanian Conservation Trust, 102 Bathurst Street, Hobart, Tasmania, 7000, or Wildlife Division, Department of Lands, Parks & Wildlife, 134 Macquarie Street, Hobart, Tasmania, 7000. In particular sightings of species additional to those presented in Appendix 11 would be welcomed. Thankyou.

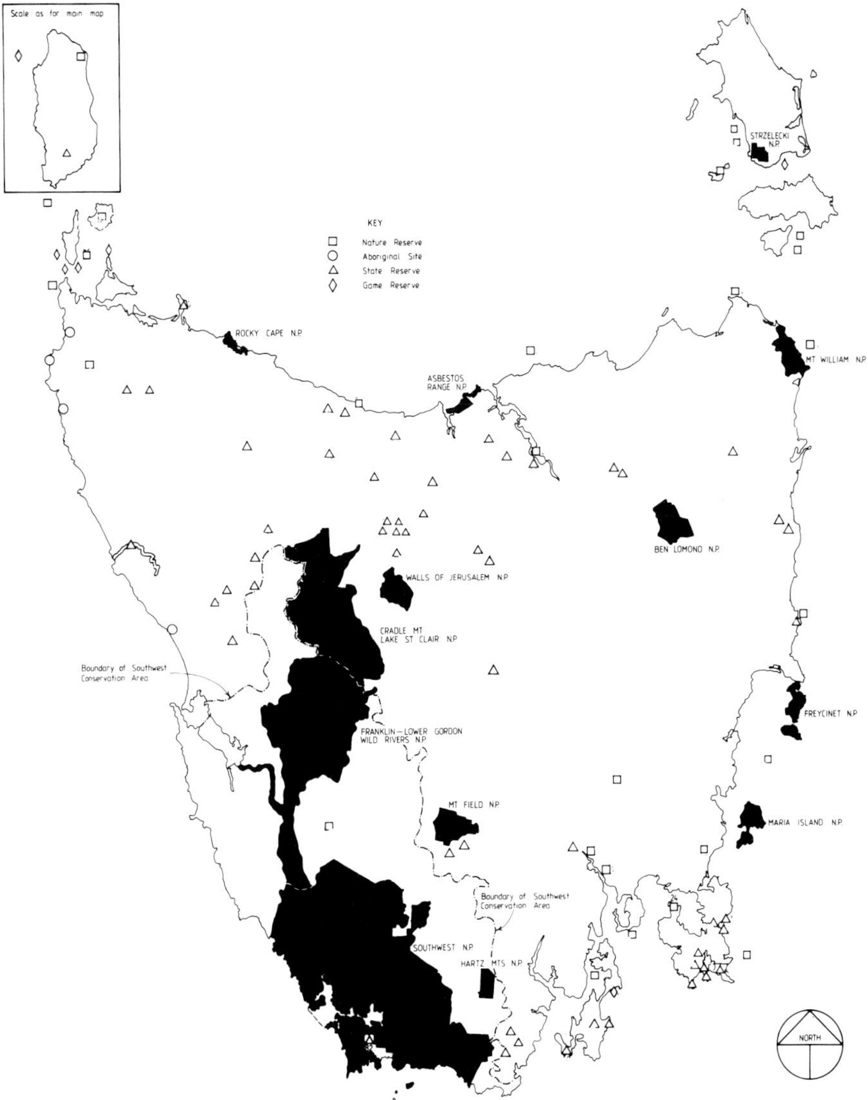

Map courtesy of Department of Lands, Parks & Wildlife

INDEX